The Human Cycle

by Nina Morgan

Illustrations by John Yates

Thomson Learning • New York

Titles in the series

The Human Cycle
The Food Cycle
The Plant Cycle
The Water Cycle

Words printed in **bold** can be found in the glossary on page 30.

First published in the
United States in 1993 by
Thomson Learning
115 Fifth Avenue
New York, NY 10003

First published in 1993 by
Wayland (Publishers) Ltd.

Library of Congress Cataloging-in-Publication Data
applied for

ISBN 1-56847-094-0

Printed in Italy

Contents

The human cycle

OPPOSITE   These three pictures are of James at different times in his life. How has he changed since he was a baby?

BELOW   The children in this group are about the same age. They probably have a lot in common. Maybe they go to the same school. They might even be good friends. Yet every single one of them is different. No two people on this earth are exactly the same.

There are billions of people in the world, but there is no one on earth exactly like you. All human beings are different. Yet every single person comes into and leaves the world in the same way. We are born, we grow up, and perhaps have children of our own one day. Eventually we die. This is known as the human life cycle.

Look at some old photographs of yourself when you were a baby. How have you changed since then?

You probably have a lot more hair now. You will certainly have grown. These changes are a natural part of growing up and the human life cycle.

As we get older, we carry on growing and changing. Our bodies get ready for the time when we can have children of our own.

Even when we grow up and have children, we continue to change. Day by day we get a little older, until eventually we become elderly. This book is about the changes that happen to us during our lifetimes.

See for yourself

Our faces change quite a lot as we grow older. Ask some friends to show you pictures of themselves as babies. Pin all the pictures on a wall. Now, try to guess which picture belongs to which friend.

A new life begins

Like all human life, your life began with one **cell**. During sexual intercourse, a single **sperm cell** unites with a single egg cell, or **ovum**. Sperm cells leave the male **penis** and enter the female **vagina**. They swim toward the **womb**, or **uterus**, where one sperm cell joins a waiting egg cell to make one new cell. This process is called **fertilization**.

The newly created cell divides to make many more cells that eventually become a fully developed baby.

Fertilization happens inside the mother's uterus. The sperm cell and the egg cell are very tiny, so we cannot see fertilization happening without a microscope. These pictures show what happens.

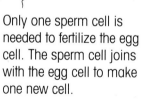

Only one sperm cell is needed to fertilize the egg cell. The sperm cell joins with the egg cell to make one new cell.

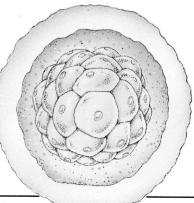

The new cell begins to divide to make a small ball of cells, which settles in the wall of the uterus.

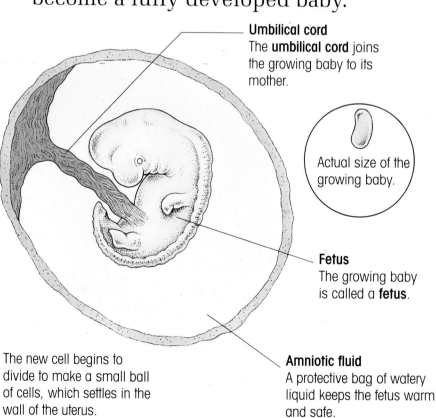

Umbilical cord
The **umbilical cord** joins the growing baby to its mother.

Actual size of the growing baby.

Fetus
The growing baby is called a **fetus**.

Amniotic fluid
A protective bag of watery liquid keeps the fetus warm and safe.

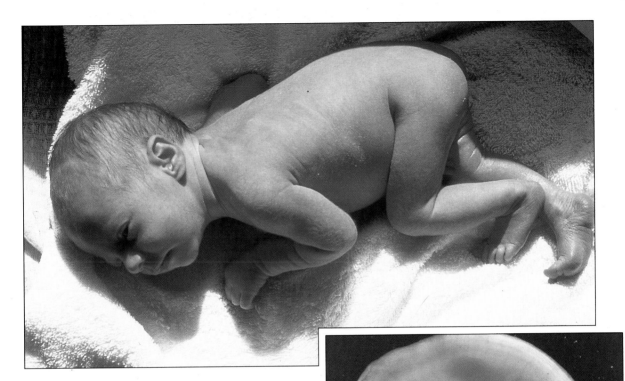

When a baby is growing inside a woman we say that she is pregnant. As you grew inside your mother, you were attached to her by the umbilical cord. The umbilical cord has a special job. A growing baby (fetus) has to get the **nutrients** it needs to grow from its mother's blood. The blood travels between mother and baby through the umbilical cord. A bag of liquid forms around the baby to protect it.

At first you looked a little like a tadpole. After only two months of

The top picture shows a healthy newborn baby. Compare it to the picture underneath of a baby growing in its mother's uterus. This fetus is only four weeks old.

growing you began to look like a tiny human being. You were still only one inch long, but already you had all your fingers and toes. After only five months, your mother could feel you move and kick inside her. At the hospital doctors use an **ultrasound scanner** to produce a moving picture of you on a screen.

By the seventh month you were formed, with eyelashes and fingernails.

The diagram below shows inside a pregnant woman. The doctor in the picture cannot see inside the woman without the help of very special equipment. However, he can tell which way the baby is lying in the uterus by feeling the mother's stomach.

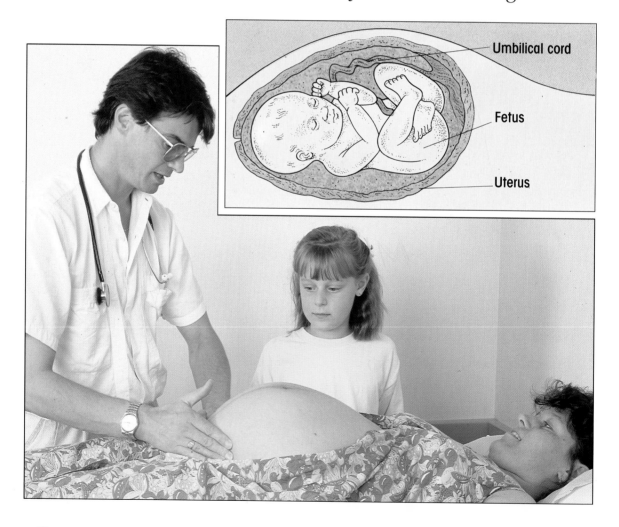

Umbilical cord

Fetus

Uterus

Sometimes there are problems. Some babies are born too early, before they are able to survive outside their mothers. These early, or premature, babies need help to survive. Like the baby in the picture above, they are placed in an **incubator** at the hospital. Incubators help to keep them alive, until they are able to survive by themselves.

This baby is in an incubator. He was born too early and so needs special help to survive. He is safe and warm inside the incubator.

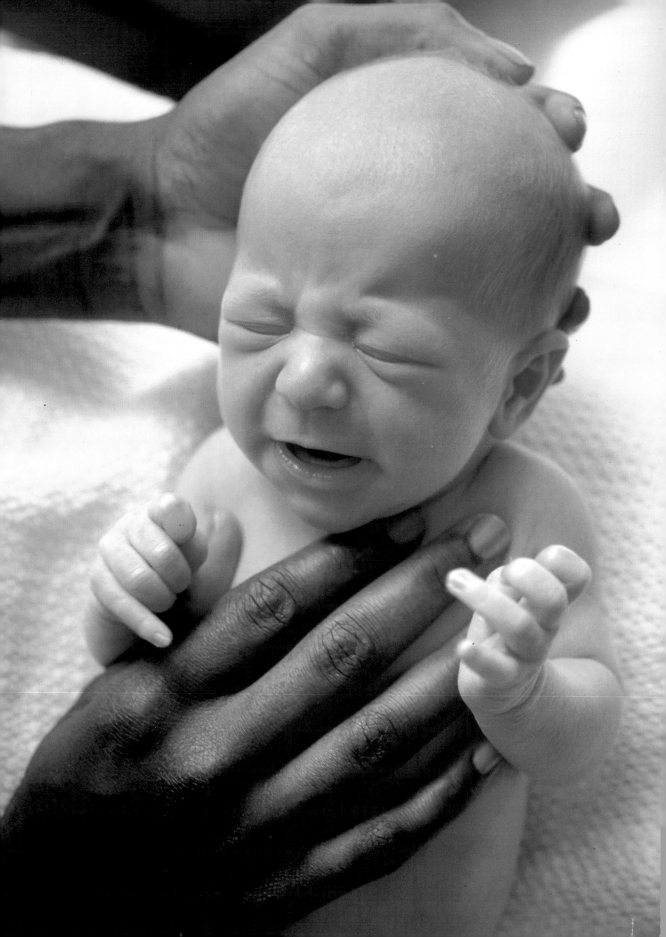

After nine months you were ready to be born. When a baby is born, its head moves downward. Muscles inside the mother push out the baby, usually head first, through the opening between her legs called the vagina.

When you were born, people probably tried to decide who you looked like. The truth is that babies always look a little like each of their parents. This is because a baby is made from an egg cell from its mother and a sperm cell from its father. Brothers and sisters often look alike too, if they have the same father and mother.

The first year

When you were first born you looked quite helpless. Even so, you were already able to do many things. Almost as soon as you were born you were able to breathe in air. You could cry loudly to attract attention and knew how to suck to get food. Although you could not see clearly, you could make out people and toys when they were placed in front of you.

In the first year of your life, you grew very quickly and learned all kinds of very important things. By three months old you could smile and gurgle, and your eyes could follow a

Newborn babies need a lot of love and care. They are unable to take care of themselves. Very young babies cannot even sit up.

This mother is breast-feeding her baby. The mother's milk contains ingredients that nourish the baby and help to protect it from illness.

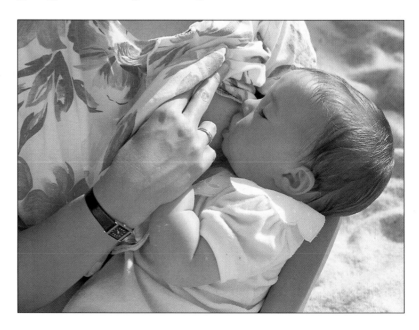

moving object. After six months you may have been able to sit up. You probably used your hands to hold things. At ten months many babies can move their arms and legs well enough to crawl. By the time you were one year old, you may even have learned to walk and could explore new things.

However, there were still things you could not yet do. You needed someone to bathe and dress you, to prepare your meals, and to feed you. All this care and attention was provided by adults, who can help babies learn about the world around them while protecting them.

This pale yellow robin is feeding its chick a bug. Even in the animal world, many babies are cared for by their parents until they are old enough to survive by themselves.

Growing up

An important part of growing up is learning new skills and developing our ability to think. This boy is learning to control his hand movements as he paints a picture.

As you get older, you grow more slowly. Usually by the time they are twenty, people stop growing altogether. However, *you* still have a lot of growing to do before you reach your full height. Why not measure yourself every two or three months? You will be able to see just how much and how quickly you grow.

As you grow taller, your feet grow bigger too. When you visit a shoe store to buy new shoes, you need to have your feet measured. This tells you if your feet have grown.

You don't just get bigger when you grow – the shape of your body changes too. If you compare the shape of your body with the shape of a baby's body, you will see many differences. Babies have short legs and big heads.

As you get older the shape of your body changes too. Younger children, like the little boy in this picture, have large heads. His sister's head is about the same size, even though she is much older and taller.

See for yourself

Everyone grows at a different speed. Measure the height of your friends. It is surprising how children of the same age will differ in size.

Children's legs are much longer, while their heads do not grow much. This means that their heads look small in comparison to the rest of their bodies. The shape of your face changes when you lose your baby teeth and grow new, adult teeth.

It is not only our bodies which change as we grow up. Our personalities change as we learn more about the world. Gradually, we learn how to take better care of ourselves and take more responsibility for our lives.

Even quite young children can take some responsibility for their lives. These two girls are tidying the bedroom after they have made a mess.

Puberty

Between the ages of ten and sixteen, changes inside our bodies start to make it possible for us to have children of our own one day. This time of change is called **puberty**, when girls and boys start to become women and men.

One of the first things that girls and boys notice is that their bodies gradually begin to change shape. Girls grow **breasts** and their hips get wider. Boys' shoulders become wider.

Between the ages of ten and sixteen, changes start to happen to our bodies. This time of change is called puberty. It means that our bodies are starting to prepare for the time when we can have children of our own.

Both boys and girls begin to grow hair. Underarm hair and pubic hair, which grows between their legs, appear. Boys grow more hair on their top lips, and their voices get deeper.

One day the children in this picture will have similar shapes to their parents. The changes that happen to their bodies during puberty will enable them to have children of their own.

The body changes that take place during puberty happen to different people at different times. Everybody develops at their own speed.

One of the most important changes to happen to girls during puberty is the start of their periods. You may already have heard of periods, but you might not know exactly what they are. A girl's body contains a lifetime supply of egg cells inside her **ovaries**, which release one egg cell every month.

The egg cell travels down into her uterus. If it is not fertilized, it will leave the body through the vagina, along with a little blood – a few tablespoons in total, although it looks like more. This monthly cycle is known as **menstruation**.

The female reproductive organs

This diagram explains what happens
during menstruation.

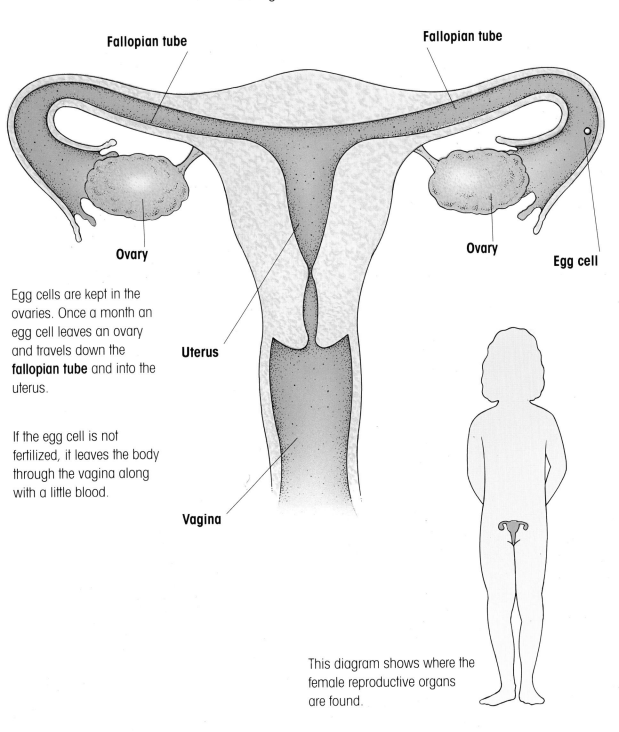

Fallopian tube

Fallopian tube

Ovary

Ovary

Egg cell

Uterus

Vagina

Egg cells are kept in the
ovaries. Once a month an
egg cell leaves an ovary
and travels down the
fallopian tube and into the
uterus.

If the egg cell is not
fertilized, it leaves the body
through the vagina along
with a little blood.

This diagram shows where the
female reproductive organs
are found.

Getting older

As we get older our bodies change in other ways. These changes take place at different times in different people.

Some changes are obvious. As we grow old our hair turns gray, and some people go bald. Our skin gets thinner and becomes wrinkled.

This man is seventy years old. Look carefully at the picture. Can you spot any signs of aging? Which parts of his appearance will not change as he gets older?

As some people get older their joints start to stiffen, so that they have difficulty walking. This man uses a zimmer frame (the metal frame in front of him) to support him as he walks.

Other changes happen inside our bodies, which we cannot see easily. As we get older there is less water in our bodies, which is why we get wrinkles. Our muscles get smaller too, especially if we do not exercise them.

Women stop menstruating around the age of 50, and they can no longer have babies. Men can still become fathers when they are quite old.

Between the ages of 60 and 70 people's bones weaken and their **joints** stiffen. During this time people also become two to three inches shorter because their backbones shrink.

Elderly people often suffer from memory loss. They can become confused and slower to react to things because their vision and hearing are no longer sharp. The body temperature of older people gradually decreases, making it more difficult to stay warm.

Many people, however, enjoy being healthy and fit in their later years, especially those who continue to eat a healthy diet and to exercise regularly.

Most working people retire around the age of 60. As a result, they have more free time to spend with their families and to enjoy their interests.

This little girl's grandfather is teaching her about gardening. We can learn a lot from older people. After all, they have lived for much longer than we have.

It has been proved that people who exercise regularly and get plenty of fresh air are more likely to remain alert and healthy until the end of their lives.

Why do we die?

This woman is 92 years old. Very old people feel cold easily and need to be kept warm.

People live for different lengths of time. Some people live for more than 100 years. Others live for only a very short time. Some babies die even before they are born. Death is a normal part of the human cycle. But why do we die?

As people age, their bodies gradually slow down and become weaker until they can no longer be maintained. This is when many people die. But people die for other reasons, too. Some die when they are young because parts of their bodies stop working properly or because they are ill. People also die in accidents. In parts of the world where there is not enough food and clean water, people can die of hunger and thirst.

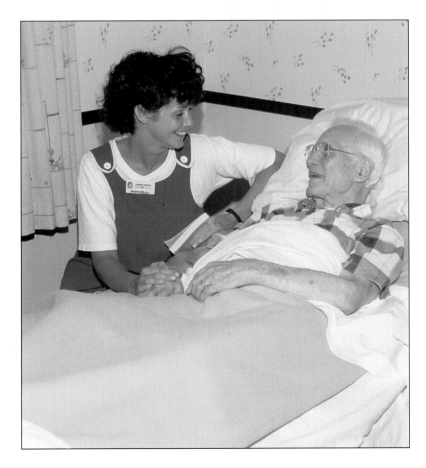

People are not always able to take care of themselves when they get old. Sometimes they have to go into nursing homes or hospitals, where they can receive special care.

These women are taking part in a funeral ceremony in Bali, Indonesia. For them, the funeral is not an unhappy occasion. They believe that before a person's soul can reach God, it must be reborn again many times in different bodies. They believe when people die, they take a step closer to God.

Doctors can often help people who are ill or injured get better. But sometimes people are too ill or too badly hurt to recover.

When someone dies, family and friends usually feel very sad. Death can be very difficult for people to accept because they miss the person who is gone. Often there is a final ceremony or funeral for the person who has died. People can show their sadness and comfort each other.

See for yourself

Cemeteries can tell us a lot about how long people live. Visit one with a teacher or other grown-up. Use the birth and death dates on gravestones to figure out how long people lived. Did most people die when they were very old? Did some people die when they were children or babies? Did women live longer than men?

Glossary

Amniotic fluid The watery liquid surrounding a developing fetus that keeps it warm and safe.

Breasts Parts of the female body that are used to feed babies. Breasts start to grow during puberty.

Cell Living units that form the basis of all living things. Cells are too tiny to be seen without a microscope. The body consists of many different kinds of cells, each with its own special functions.

Cemetery A place where people are buried after they die.

Fallopian tube A narrow tube leading from the ovaries to the uterus through which egg cells pass. Female bodies have two fallopian tubes connecting each ovary to the uterus.

Fertilization The joining of a male sperm cell with a female egg cell to form a new cell.

Fetus The name given to a developing baby inside the uterus.

Incubator A special machine used to help babies that are born too early to survive by keeping them safe and warm.

Joints Parts of the body where two bones are held together, such as the knee.

Menstruation The passing of an unfertilized egg, along with a little blood, from the uterus through the vagina. This is also called a period.

Nutrients The things our bodies need from food to stay healthy.

Ovaries Two glands inside the female body where eggs are produced.

Ovum   An egg cell.

Penis   The part the male body that is used during sexual intercourse. It is also used for urinating.

Puberty   The process usually between the ages of 10 and 16 when human bodies begin to change.

Sperm cells   The male sex cell. One sperm cell and one egg cell are needed to make a baby.

Ultrasound scanner   A machine that shows a baby in the womb.

Umbilical cord   The cord that connects the fetus to its mother's body.

Uterus   The female organ where the fetus will grow until birth. Also called the womb.

Vagina   The female passageway through which sperm cells travel to reach the egg cell. If the cell does not get fertilized, it passes through the vagina and out of the body.

Womb   Another name for the uterus.

Further Reading

Bell, Neill. *Only Human: Why We Are the Way We Are.* New York: Little, Brown, 1983.

Berger, Gilda. *The Human Body.* New York: Doubleday, 1989.

Gamlin, Linda. *The Human Body.* Today's World. New York: Gloucester Press, 1988.

Gamlin, Linda. *The Human Race.* Today's World. New York: Gloucester Press, 1988.

Peacock, Graham and Hudson, Terry. *The Super Science Book of Our Bodies.* Super Science. New York: Thomson Learning, 1993.

Prot, Vivian A. *The Story of Birth.* Ossining, NY: Young Discovery Library, 1991.

Picture acknowledgments
Images 12, 18, 24, 25; Life File 15, 26; Oxford Scientific Films 13; Science Photo Library 6, 7, 8; Skjold 23, 27, 29; Tony Stone 4, 5, 9, 10, 11, 20; Jennie Woodcock 12; Tim Woodcock 5 (all); WPL 16, 17, 22. The artwork on page 15 by Alex Pang.

Index

DATE DUE